SCHOLASTIC

Success With

Math

D1315178

New York • Toronto • London • Auckland • Sydney
Mexico City • New Delhi • Hong Kong • Buenos Aires

Teaching *Resources*

About the Book

"Nothing succeeds like success."
—Alexandre Dumas the Elder, 1854

And no other math resource helps kids succeed like Scholastic Success With Math! For classroom or at-home use, this exciting series for kids in grades 1 through 5 provides invaluable reinforcement and practice for math skills such as:

❑ number sense and concepts
❑ reasoning and logic
❑ basic operations and computations
❑ story problems and equations
❑ time, money, and measurement
❑ fractions, decimals, and percentages
❑ geometry and basic shapes
❑ graphs, charts, tables . . . and more!

Each 64-page book contains loads of challenging puzzles, inviting games, and clever practice pages to keep kids delighted and excited as they strengthen their basic math skills.

What makes *Scholastic Success With Math* so solid?

Each practice page in the series reinforces a specific, age-appropriate skill as outlined in one or more of the following standardized tests:

◎ *Iowa Tests of Basic Skills*
◎ *California Tests of Basic Skills*
◎ *California Achievement Test*
◎ *Metropolitan Achievement Test*
◎ *Stanford Achievement Test*

These are the skills that help kids succeed in daily math work and on standardized achievement tests. And the handy Instant Skills Index at the back of every book helps you succeed in zeroing in on the skills your kids need most!

Take the lead and help kids succeed with *Scholastic Success With Math*.
Parents and teachers agree: No one helps kids succeed like Scholastic!

Comparing & Ordering Numbers

Name _____ Date _____

Use the digits in the box to answer each number riddle.
You cannot repeat digits within a number.

$$1 \quad 8 \quad 3 \quad 4 \quad 9 \quad 6 \quad 2 \quad 7$$

1 I am the number that is 100 greater than 3,362.
What number am I? _____

2 I am the number that is 40 less than the largest number
you can make using five of the digits.
What number am I? _____

3 I am the largest number you can make that is greater than
8,745 but less than 8,750.
What number am I? _____

4 I am the number that is 5,000 greater than the smallest
number you can make using six of the digits.
What number am I? _____

5 I am the smallest number you can make that is greater
than 617,500.
What number am I? _____

6 I am the largest number you can make that is less than
618,400 but greater than 618,300.
What number am I? _____

Sign It!

Name _____ Date _____

✏ There are signs with numbers on them almost everywhere you look! They're on street corners and on highways. What if those numbers were written out as words?

Take a look at the street signs below. They all have numbers on them. Each sign has a blank sign next to it. Write the numbers as words on each blank sign. We've done the first one for you.

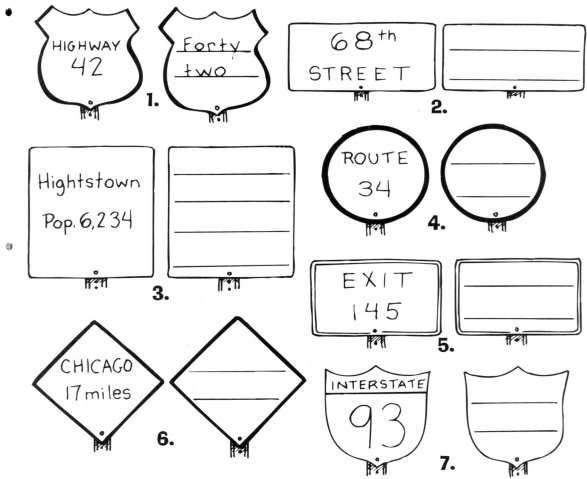

HIGHWAY 42

Forty two

1.

68th STREET

2.

Hightstown Pop. 6,234

3.

ROUTE 34

4.

EXIT 145

5.

CHICAGO 17 miles

6.

INTERSTATE 93

7.

IT'S YOUR TURN

Try writing other numbers as words, such as your address, area code, age, or shoe size.

Mystery Number

Name _____ Date _____

Use the digits in the box to answer each number riddle. Digits appear only once in an answer. Each answer may not use all digits.

2 4 9 6 7 3

1 When you subtract a 2-digit number from a 3-digit number, the difference is 473.
What are the numbers? _____

2 The sum of these two numbers is 112.
What are the numbers? _____

3 The sum of these two numbers is 519.
What are the numbers? _____

4 The difference between these two 3-digit numbers is 263.
What are the numbers? _____

5 The sum of these three 2-digit numbers is 184.
What are the numbers? _____

6 The difference between two 3-digit numbers is a palindrome between 200 and 300.
What are the numbers? _____

What Number Am I?

Name _____ Date _____

Use the digits in the box to answer each number riddle.
You cannot repeat digits within a number.

$$1\quad 2\quad 3\quad 4\quad 5\quad 6\quad 7\quad 8\quad 9$$

1 I am the largest 4-digit odd number you can make.
What number am I? _____

2 I am the smallest 5-digit even number you can make.
What number am I? _____

3 I am the largest 5-digit even number you can make that has a 3 in the thousands place.
What number am I? _____

4 I am the smallest 5-digit number you can make that has all odd digits.
What number am I? _____

5 I am the largest 6-digit number you can make that has a 1 in the thousands place and a 5 in the ten-thousands place.
What number am I? _____

6 I am the smallest 6-digit even number you can make that has a 6 in the hundreds place.
What number am I? _____

A Place for Every Number

Name _____ Date _____

Look at the numbers in 243. Each number in the group has its own "place" and meaning. For instance, the 2 in 243 is in the hundreds place. That stands for 2 hundreds or 200. The 4 is in the tens place, meaning 4 tens or 40. And the 3 is in the ones place, meaning 3 ones or 3.

DIRECTIONS:
Use a place value chart to put the numbers in this crossnumber puzzle in their places.

ACROSS
A. 3 hundreds 2 tens 6 ones
C. 8 tens 1 one
E. 6 tens 4 ones
F. 4 tens 7 ones
H. 5 hundreds 2 tens 6 ones
J. 9 tens 3 ones
K. 8 tens 9 ones
M. 5 hundreds 4 tens 2 ones
O. 2 thousands 8 hundreds 3 tens 1 one
Q. 9 tens 8 ones
S. 6 hundreds 6 tens 4 ones

DOWN
A. 3 tens 6 ones
B. 2 thousands 4 hundreds 5 tens 7 ones
D. 1 hundred 4 tens 9 ones
G. 7 tens 3 ones
I. 6 thousands 8 hundreds 5 tens 1 one
L. 9 tens 4 ones
N. 2 hundreds 9 tens 6 ones
P. 3 tens 5
R. 8 tens 4 ones

Make a list of 10 numbers written out in the same way as the clues above. Ask a classmate to write each of those numbers in their own place value box.

9

Bee Riddle

Name _____ Date _____

Riddle: What did the farmer get when he tried to reach the beehive?

Round each number. Then use the Decoder to solve the riddle by filling in the spaces at the bottom of the page.

Decoder

400 .... A
800 .... W
30 .... O
10 .... Y
25 .... E
500 I
210 .... J
20 L
40 C
700 .... U
90 .... S
100 .... T
600 .... G
95 .... F
50 .... N
550 V
300 .... Z
7 .... H
200 .... Z

❶ Round 7 to the nearest ten _____

❷ Round 23 to the nearest ten _____

❸ Round 46 to the nearest ten _____

❹ Round 92 to the nearest ten _____

❺ Round 203 to the nearest hundred _____

❻ Round 420 to the nearest hundred _____

❼ Round 588 to the nearest hundred _____

❽ Round 312 to the nearest hundred _____

❾ Round 549 to the nearest hundred _____

❿ Round 710 to the nearest hundred _____

A "B __ __ __ __ " __ __ __ __ __ __
    10   5   8   1   4   9   7   3   6   2

When to Estimate

Name _____ Date _____

Estimation is a great way to solve many problems.

But some problems need an exact answer. How can you decide?

Read each question below. Think about what kind of answer you need.
Then circle Estimate or Exact Answer.

1. How much sugar do you need to make cookies? Estimate Exact Answer

2. How much money could your school play earn? Estimate Exact Answer

3. How many plates will you need to serve dinner? Estimate Exact Answer

4. How much money will three new tapes cost? Estimate Exact Answer

5. How long will it take to get to the airport? Estimate Exact Answer

6. How much money is in a bank account? Estimate Exact Answer

7. How long would it take you to run a mile? Estimate Exact Answer

8. How many kids are in your class? Estimate Exact Answer

How Would You Estimate . . .

On another sheet of paper, write about how you would estimate each of these.

...the height of a tree?

...how long it would take to walk from Miami to Seattle?

...how much water you use in a year?

...the number of gumballs in a gumball machine?

...the number of students in your school?

...how much one million pennies would weigh?

Super Seven

Name _____ Date _____

How can you make the number seven even?

Find the answer by completing the next step in the pattern. Then use the Decoder to solve the riddle by filling in the blanks at the bottom of the page.

Decoder

5	B
1	A
97	D
215	Y
22	W
124	H
31	I
2	P
115	A
120	C
50	N
4	E
60	S
232	M
26	T
100	R
32	F
57	E
34	K

❶ 10, 7, 4, ___

❷ 19, 13, 8, ___

❸ 42, 40, 36, 30, ___

❹ 56, 54, 50, 42, ___

❺ 33, 32, 34, 33, 35, ___

❻ 117, 97, 77, ___

❼ 205, 175, 150, 130, ___

❽ 344, 274, 214, 164, ___

❾ 760, 660, 540, 400, 240, ___

❿ 512, 490, 457, 413, 358, 292, ___

TA __ __   __ __ __   __ __ __ __   " __ ".
   5 2   7 3 1 10 4 8 6   9

Pansy's Picture Patterns

Name _____ Date _____

✎ Pansy Pattern has lots of hobbies. Her favorite hobby, though, is drawing patterns. There's just one problem. Sometimes Pansy forgets to draw the complete pattern. Maybe you can help. Try filling in the missing pieces in the patterns below.

1. 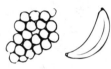  _____ _____

2. _____   _____ _____ _____

3.  _____   _____

4.    _____

5.    _____

 Draw a picture pattern of your own. Ask a classmate to fill in the missing pictures.

We All Need Friends

Name _____ Date _____

Riddle: What is a ten without its number-one friend?

Find the answers. Then use the Decoder to solve the riddle by filling in the blanks at the bottom of the page.

❶ Joe has 5 apples. Sam has 11. How many apples do both Joe and Sam have? _____

❷ Heather has 20 apples. Gina has 9. How many more apples does Heather have than Gina? _____

❸ Gary has 8 more apples than Steve. Steve has 32 apples. How many apples does Gary have? _____

❹ Brenda has 25 apples. If she gives 19 apples to Julie, how many will she have left? _____

❺ Jack has 4 apples, 7 oranges, and 9 bananas. Susan has 8 oranges, 2 bananas, and 10 apples. How many apples do both Jack and Susan have? _____

❻ Jim has 23 apples, 16 pears, and 4 bananas. Holly has 8 pears, 13 bananas, and 21 apples. Who has more fruit? _____

❼ Debbie has half the number of oranges that Patti has. Patti has 34 oranges. How many oranges does Debbie have? _____

❽ Alan has double the number of pears that Dave has. Dave has 9 pears. How many pears does Alan have? _____

❾ Beth has 50 bananas. She gives 5 to Kathy, 12 to Jeff, and 8 to Nadine. How many bananas does Beth have left? _____

❿ Zachary has 100 apples. He gives 10 to Jeremy. Then he gives half of his remaining apples to Maria. How many apples does Maria get? _____

Decoder

Holly C
24 apples .... F
9 apples .... U
18 oranges I
20 oranges .... K
16 apples .... T
25 apples .... M
18 pears .... O
50 apples .... S
17 oranges .... Z
45 apples .... A
14 apples .... R
6 apples .... A
5 apples .... R
25 bananas .... T
Jim .... O
40 apples .... E
32 pears .... W
11 apples L

___ ___ ___ ___ ___ ___   ___ ___ ___ ___
 4   1   6   9 10   2   7   3   5   8

Food to Go

Name _____ Date _____

Figure it out!

1. Woovis the Dog found a $5 bill. Which item can he buy from the menu that will give the least change?

_____

DOGGIE DINER BREAKFAST SHOP

MENU

Kibble $1.79
Mouse Crumbs $1.39
Table Scraps $2.49
Crumbs & Cheese $3.19
Deluxe Scraps $3.79

2. Molly Mouse gets Crumbs & Cheese for breakfast. She pays with the $5 bill. With the leftover money, what can Woovis buy to eat?

_____

3. Which item can Woovis buy with the $5 bill that will give the most change?

_____

4. Which two items can Woovis buy with the $5 bill so that he gets about $1 back in change?

_____

5. Woovis ordered two items from the menu and gave the cashier the $5 bill. But the two items cost more than $6.50. Which two items did Woovis order?

_____

SUPER CHALLENGE: Can Woovis use the $5 bill to buy three different items from the menu? Why or why not?

15

Root for the Home Team!

Name _____   Date _____

Riddle: What do cheerleaders like to drink?

Use the coordinates to identify points on the graph. Then use the point names to solve the riddle by filling in the blanks at the bottom of the page.

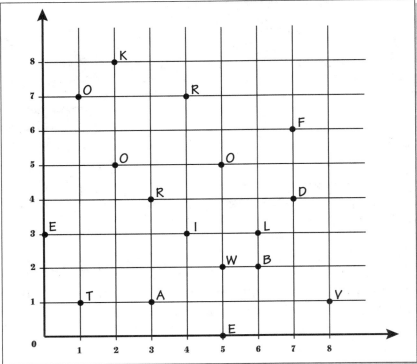

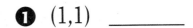

❶ (1,1) _____   ❻ (2,5) _____

❷ (3,4) _____   ❼ (0,3) _____

❸ (4,7) _____   ❽ (1,7) _____

❹ (6,2) _____   ❾ (7,6) _____

❺ (5,5) _____   ❿ (5,0) _____

LOTS ___ ___   ___ ___ ___ ___   ___ ___ ___ ___
    5   9   2 6 8 1   4 7 10 3

Horseplay

Name _____ Date _____

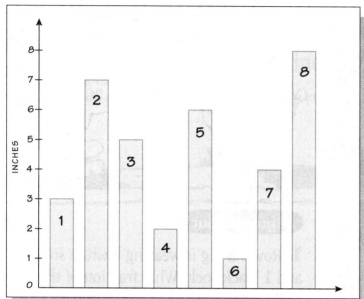

Why did the horse sneeze?

KACHOO!

Decoder

4 bars .... **T**

6 inches .... **K**

bar 5..................... **L**

bar 2..................... **A**

bar 6..................... **L**

2 inches .... **U**

2 bars .... **P**

5 inches .... **L**

bar 8..................... **T**

7 inches .... **W**

bar 3..................... **O**

bar 1..................... **S**

3 inches .... **E**

8 inches .... **C**

8 bars .... **M**

4 inches .... **T**

3 bars .... **H**

bar 4 .... **N**

bar 7..................... **I**

Answer each question about the graph. Then use the Decoder to solve the riddle by filling in the blanks at the bottom of the page.

❶ Which is the tallest bar on the graph? _____

❷ Which is the shortest bar on the graph? _____

❸ How tall is bar 1? _____

❹ How much taller is bar 5 than bar 4? _____

❺ How much shorter is bar 4 than bar 2? _____

❻ How tall is bar 8? _____

❼ Which bar is taller: bar 1 or bar 7? _____

❽ Which bar is shorter: bar 2 or bar 3? _____

❾ Which bar is twice the size of bar 1? _____

❿ How many of bar 4 would equal bar 8? _____

IT HAD A __ __ __ __ __ __ " __ __ __."
2 7 10 4 9 3   6 8 5 1

A Sick Riddle

Name _____   Date _____

Riddle: What sickness can't you talk about until it's cured?

Find each sum. Then use the Decoder to solve the riddle by filling in the spaces at the bottom of the page.

❶ 12 + 7   = _____

❷ 32 + 10   = _____

❸ 50 + 4   = _____

❹ 13 + 22   = _____

❺ 47 + 19   = _____

❻ 97 + 68   = _____

❼ 204 + 41 = _____

❽ 37 + 331 = _____

❾ 670 + 98 = _____

❿ 857 + 466 = _____

Decoder

66 I
57 .... W
42 I
216 .... M
19 .... Y
97 C
768 .... G
35 .... S
46 .... E
100 X
245 .... R
1,257 .... D
54 .... A
52 .... O
368 L
82 .... P
1,323 .... T
155 .... Q
165 .... N

____ ____ ____ ____ ____ ____ ____ ____ ____ ____
 8   3   7   1   6   9   5   10   2   4

Blooming Octagon

Name _____   Date _____

Solve the problems. ◆ If the answer is between 1 and 300, color the shape yellow. ◆ If the answer is between 301 and 600, color the shape blue. ◆ If the answer is between 601 and 1,000, color the shape orange. ◆ Finish by coloring the outer shapes with the colors of your choice.

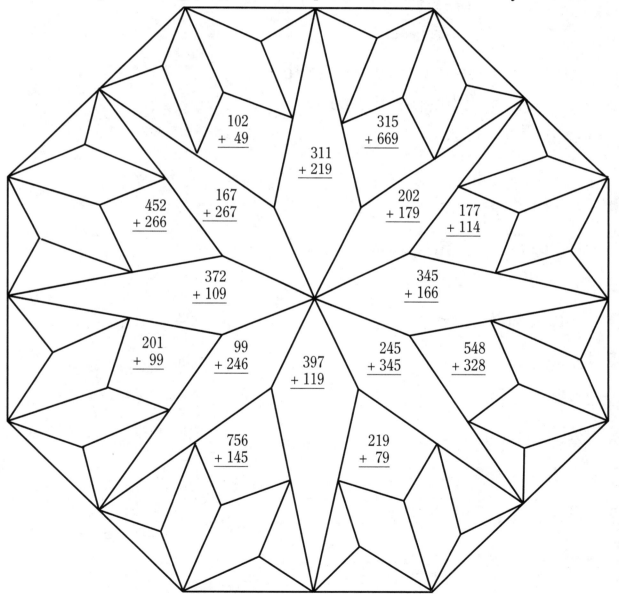

Taking It Further: Fill in the next three numbers in this pattern.

150, 300, 450, 600, _____, _____, _____.

21

Bathtub Brunch

Name _____ Date _____

MENU

Riddle: What's the best thing to eat in a bathtub?

Find each sum. Then use the Decoder to solve the riddle by filling in the spaces at the bottom of the page.

Decoder

5,429	A
10,493	F
2,133	S
14,983	R
10,439	P
712	U
3,489	K
1,840	M
1,063	E
4,523	W
689	N
2,009	B
8,292	O
3,234	I
7,538	G
1,804	C
4,708	H
6,521	L
8,234	E

1 $1,004 + 800$ = _____

2 $512 + 177$ = _____

3 $364 + 699$ = _____

4 $1,245 + 888$ = _____

5 $1,876 + 1,613$ = _____

6 $2,010 + 6,224$ = _____

7 $5,470 + 2,068$ = _____

8 $4,526 + 3,766$ = _____

9 $1,017 + 4,412$ = _____

10 $2,588 + 7,851$ = _____

‾4‾ ‾10‾ ‾8‾ ‾2‾ ‾7‾ ‾6‾   ‾1‾ ‾9‾ ‾5‾ ‾3‾

Sums & Differences

Name _____ Date _____

> **Point-scoring in the Inter-Galaxy Football League**
> Touchdown.. 6 points
> Touchdown with an extra point.................... 7 points
> Touchdown with a 2-point conversion .... 8 points
> Field Goal.. 3 points

The Asteroids played the Constellations. Each team scored a field goal in the first quarter. In the second quarter, the Asteroids scored a touchdown, but missed the extra point. At the half, the Constellations led by 1 point. In the third quarter, the Asteroids made a touchdown with the extra point. The Constellations matched them, and made a field goal, as well. In the fourth quarter, following a Constellation field goal, the Asteroids scored a touchdown with a 2-point conversion.

Who won? _____

By what score? _____

> **Point-scoring in the Inter-Galaxy Basketball League consists of
> 1-point free throws, 2-point goals, 3-point goals, and 4-point goals
> (those made without looking at the basket!).**

The Comets, playing the Meteors, led 22–9 at the end of the first quarter. They led by 7 at the half after scoring two 4-point goals, two 3-point goals, four 2-point goals, and three free throws. In the third quarter, the Meteors had six 2-point goals and four free throws. They also had one more 4-point goal, but one less 3-point goal than the Comets. The Comets had five 2-point goals and no free-throws. They scored 20 points in the quarter. In the last quarter, each team scored the same number of 4-point, 3-point, and 2-point goals. The Comets scored 31 points in that quarter, including four free throws. The Meteors made two fewer free throws than the Comets.

Who won? _____

By what score? _____

A Friendly Riddle

Name _____ Date _____

Riddle: When is the ocean friendliest?

Do each subtraction problem. Then use the Decoder to solve the riddle by filling in the spaces at the bottom of the page.

Decoder

25 .... **B**
64 .... **V**
29 .... **I**
37 .... **T**
7 .... **N**
18 .... **E**
586 .... **C**
6 .... **H**
97 .... **E**
13 .... **D**
11 .... **W**
4 .... **R**
3 .... **A**
28 .... **S**
32 .... **O**
486 .... **N**
112 .... **X**
8 .... **I**
14 .... **M**

1 10 – 4   = _____

2 19 – 16   = _____

3 35 – 7   = _____

4 27 – 19   = _____

5 50 – 32   = _____

6 84 – 47   = _____

7 117 – 20   = _____

8 155 – 144 = _____

9 314 – 250 = _____

10 572 – 86   = _____

W ___ ___ ___   ___ ___   ___ ___ ___ ___ ___.
   1   7   10   4   6   8   2   9   5   3

Box of Many Colors

Name _____ Date _____

Solve the problems. ◆ If the answer is between 0 and 200, color the shape red. ◆ If the answer is between 201 and 400, color the shape orange. ◆ If the answer is between 401 and 600, color the shape yellow. ◆ If the answer is between 601 and 800, color the shape blue. ◆ If the answer is between 801 and 999, color the shape purple.

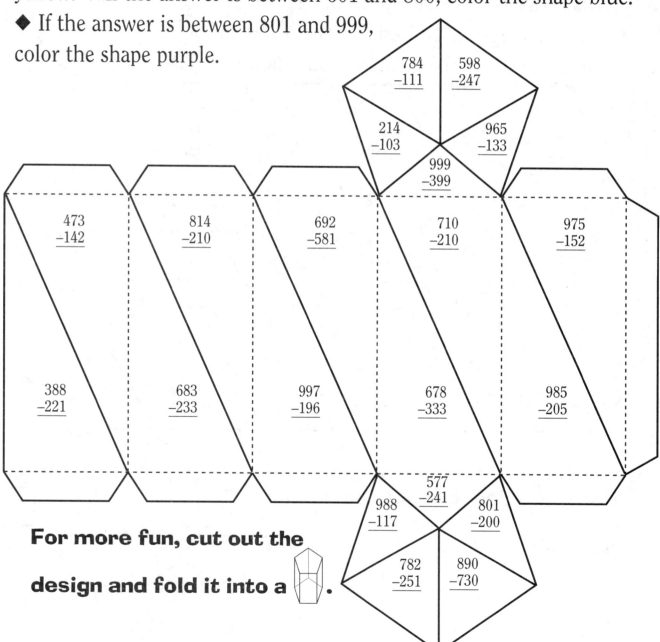

For more fun, cut out the design and fold it into a .

Gemstones

Name _____ Date _____

Solve the problems. ◆ If the answer is between 0 and 300, color the shape red. ◆ If the answer is between 301 and 600, color the shape green. ◆ If the answer is between 601 and 999, color the shape yellow. ◆ Finish the design by coloring the other shapes with the colors of your choice.

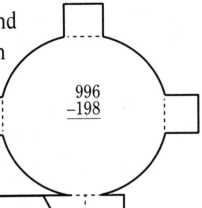

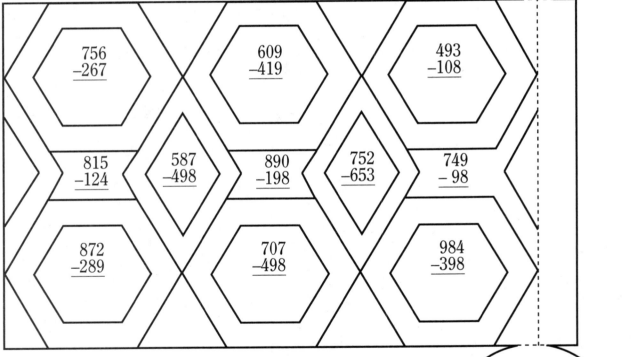

756
−267

609
−419

493
−108

815
−124

587
−498

890
−198

752
−653

749
− 98

872
−289

707
−498

984
−398

For more fun, cut out the design and fold it into a ⬚ **.**

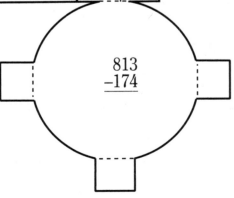

813
−174

Soccer Balls

Name _____ Date _____

Solve the problems, then choose two colors that you like. ◆ Write the name of one of the colors on each line below. ◆ Color the design. If the answer is even, color the shape _____. If the answer is odd, color the shape _____. ◆ Finish the design by coloring the other shapes with the colors of your choice.

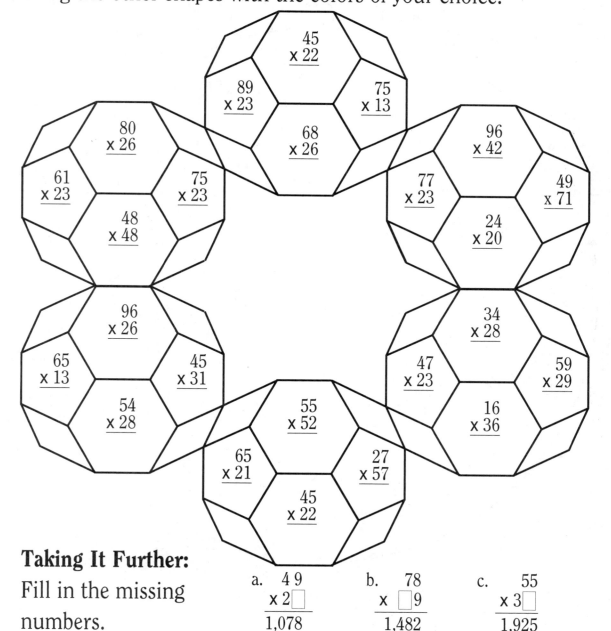

Taking It Further:
Fill in the missing numbers.

a. 49
    x 2☐
    ―――
    1,078

b. 78
    x ☐9
    ―――
    1,482

c. 55
    x 3☐
    ―――
    1,925

In the Wink of an Eye

Name _____   Date _____

Solve the problems. If the answer is even, connect the dot beside each problem to the heart on the right- and left-hand sides of the circle. If the answer is odd, do nothing. Two lines have been drawn for you.

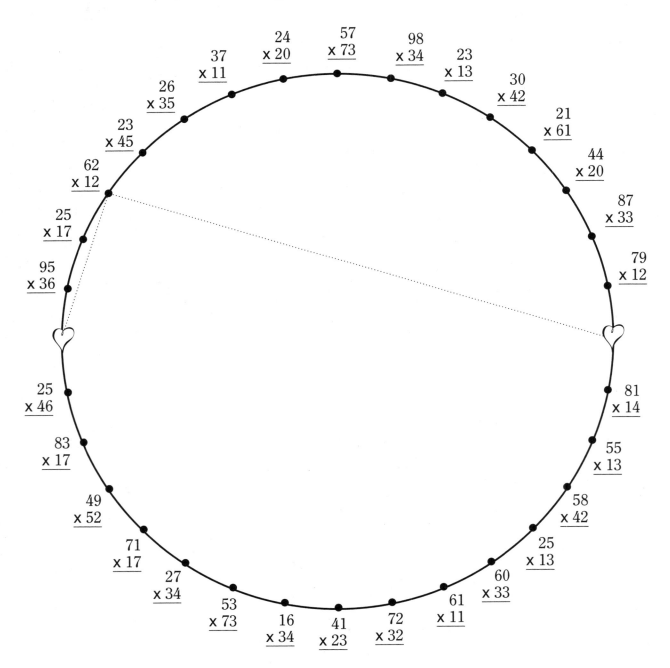

Ship Shape

Name _____ Date _____

What are the cheapest
ships to buy?

What To Do

To find the answer to the riddle,
solve the multiplication problems.
Then, match each product with a
letter in the Key below. Write the
correct letters on the blanks below.

1 100 x 23 = _____

2 200 x 17 = _____

3 300 x 31 = _____

4 400 x 44 = _____

5 500 x 19 = _____

6 600 x 27 = _____

7 700 x 35 = _____

8 800 x 18 = _____

9 900 x 50 = _____

Key

3,200 .... D	16,200 .... B	16,700 .... H
17,600 L	3,600 .... K	24,500 .... O
10,500 I	3,400 .... T	12,600 .... Y
45,000 .... A	9,300 .... E	14,400 .... A
15,300 .... R	9,500 .... S	2,300 .... S

Riddle
Answer: "_____"
        5 **9** **4** **3**   **6** **7** **8** **2** **1**

Purple Blossoms

Name _____ Date _____

Solve the problems. ◆ If the answer is between 1 and 250, color the shape yellow. ◆ If the answer is between 251 and 4000, color the shape purple. ◆ If the answer is between 4,001 and 9,000, color the shape pink. ◆ Finish by coloring the other shapes with colors of your choice.

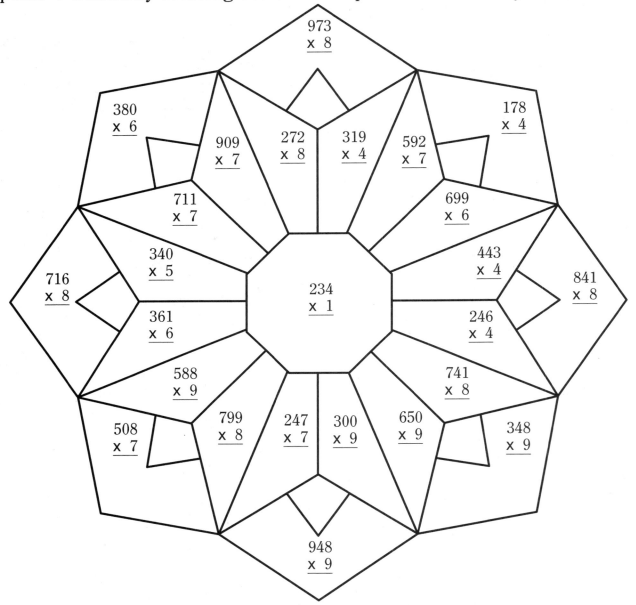

Taking It Further: I am an even number. I have three digits and they are all the same. If you multiply me by 4, all of the digits in the product are 8. What number am I? _____

Caught in the Web

Name _____ Date _____

Why did the spider join the baseball team?

What To Do

To find the answer to the riddle, solve the multiplication problems. Then, match each product with a letter in the Key below. Write the correct letters on the blanks below.

1 **1,000 x 11 =** _____

2 **2,000 x 12 =** _____

3 **3,000 x 10 =** _____

4 **4,000 x 14 =** _____

5 **5,000 x 20 =** _____

6 **6,000 x 24 =** _____

7 **7,000 x 30 =** _____

8 **8,000 x 32 =** _____

9 **9,000 x 40 =** _____

10 **7,500 x 50 =** _____

Key

56,000	H	65,000	M	30,000	C
11,000	I	144,000	T	375,000	C
265,000	B	25,000	N	10,000	Y
360,000	F	256,000	L	100,000	A
210,000	E	90,000	Q	24,000	S

Riddle Answer: **TO** ___ ___ ___ ___ ___ " ___ ___ ___ ___ ___ "
3 **5** **6** **10** **4**   **9** **8** **1** **7** **2**

Patchwork Diamonds

Name _____ Date _____

Solve the problems. ◆ If the answer is between 1 and 6, color the shape green. ◆ If the answer is between 7 and 12, color the shape red. ◆ Finish the design by coloring the other shapes with the colors of your choice.

Taking It Further: Jamie is making a quilt with 70 diamond-shaped pieces. If 7 pieces make 1 square, how many squares will her quilt have?

Triangle Patches

Name _____ Date _____

Solve the problems. ◆ If the answer is 9 or less, color the shape green.
◆ If the answer is 10 or greater, color the shape orange. ◆ Finish the
design by coloring the other shapes with the colors of your choice.

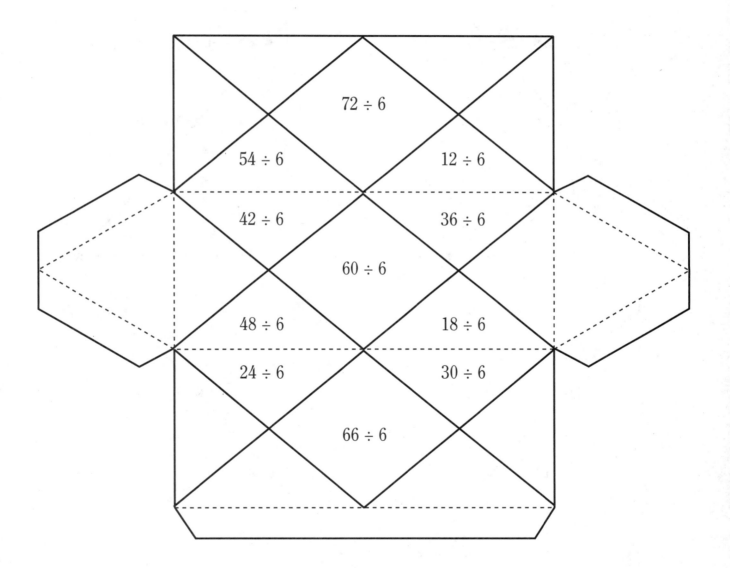

For more fun, cut out the

design and fold it into a **.**

Mirror Image

Name _____ Date _____

Solve the problems. Then connect the dot beside each problem to the dot beside its answer.

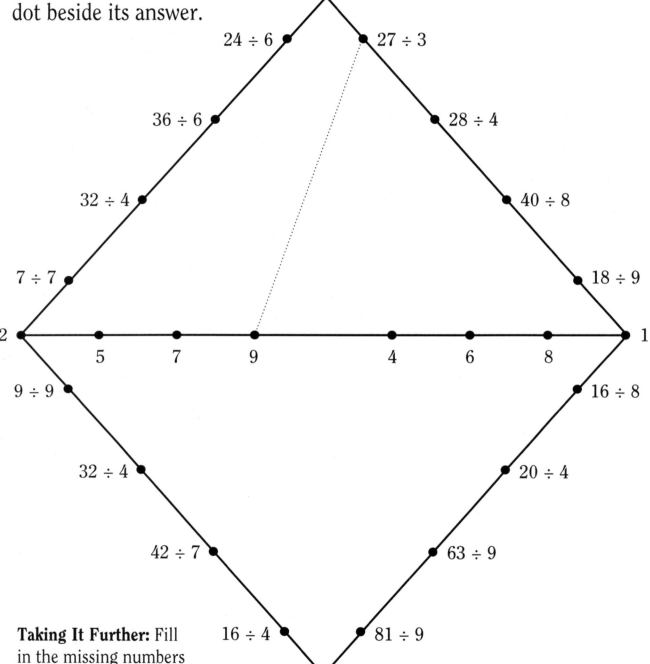

Taking It Further: Fill in the missing numbers in the patterns below.

81, 72, 63, _____, _____, _____, _____, _____, 9

63, 56, 49, _____, _____, _____, _____, _____, 7

Circus Tent

Name _____ Date _____

Solve the problems.
If the answer is even,
color the shape purple.
If the answer is odd,
color the shape red.

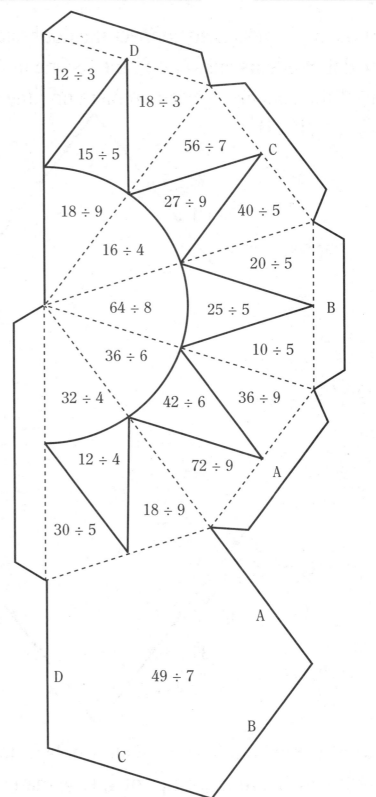

**For more fun,
cut out the
design and
fold it into a** △.

$12 \div 3$
$18 \div 3$
$15 \div 5$
$56 \div 7$
$18 \div 9$
$27 \div 9$
$40 \div 5$
$16 \div 4$
$20 \div 5$
$64 \div 8$
$25 \div 5$
$36 \div 6$
$10 \div 5$
$32 \div 4$
$42 \div 6$
$36 \div 9$
$12 \div 4$
$72 \div 9$
$30 \div 5$
$18 \div 9$
$49 \div 7$

D
C
B
A

35

Football

Name _____   Date _____

Solve the problems. Then connect the dot beside each problem on Line A
to the dot beside its answer on Line B. One line has been drawn for you.
Connect the dots beside each problem on Line C to the dot beside its
answer on Line D.

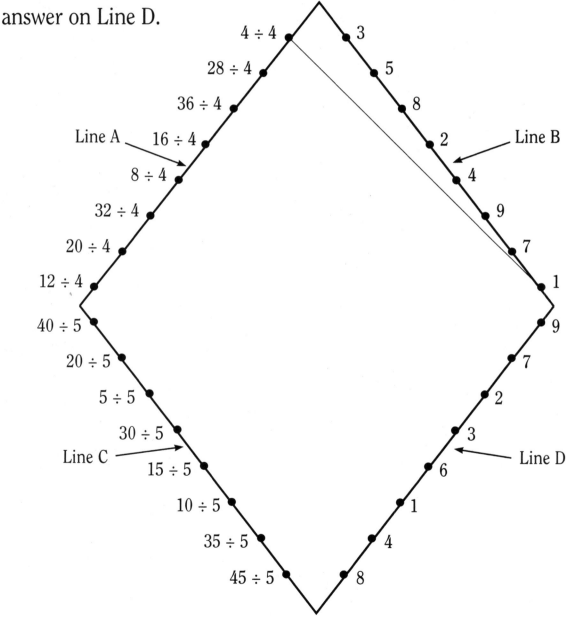

Taking It Further: There are 40 people in line at the museum. If they
tour the museum in groups of 4, how many groups will there be?

Flying Carpet

Name _____   Date _____

Solve the problems. ◆ If the answer is between 100 and 250, color the shape red. ◆ If the answer is between 251 and 900, color the shape blue. ◆ Finish the design by coloring the other shapes with the colors of your choice.

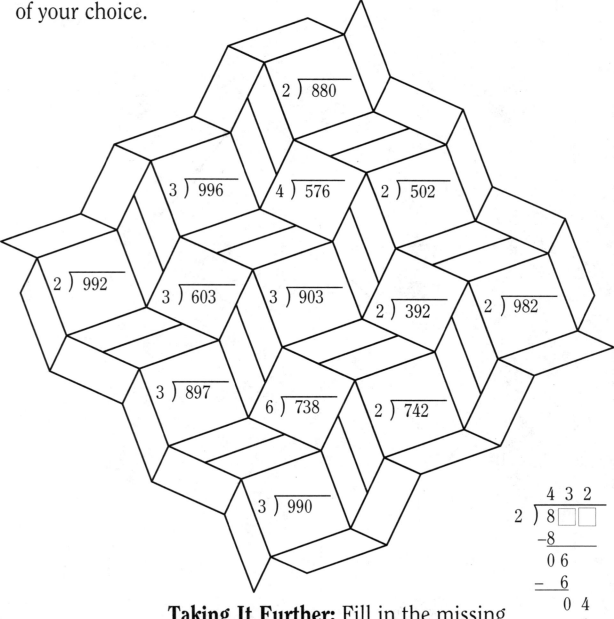

$$2\overline{)880}$$

$$3\overline{)996} \qquad 4\overline{)576} \qquad 2\overline{)502}$$

$$2\overline{)992}$$

$$3\overline{)603} \qquad 3\overline{)903} \qquad 2\overline{)392} \qquad 2\overline{)982}$$

$$3\overline{)897}$$

$$6\overline{)738} \qquad 2\overline{)742}$$

$$3\overline{)990}$$

$$\begin{array}{r} 4\ 3\ 2 \\ 2\overline{)8\square\square} \\ -8 \\ \hline 0\ 6 \\ -\ 6 \\ \hline 0\ 4 \\ -\ 4 \\ \hline 0 \end{array}$$

Taking It Further: Fill in the missing digits in the problem to the right.

37

Honeycomb

Name _____ Date _____

Solve the problems. ◆ If the answer has a remainder between 1 and 4, color the shape black. ◆ If the answer has a remainder between 5 and 8, color the shape red. ◆ Finish the design by coloring the other shapes with the colors of your choice.

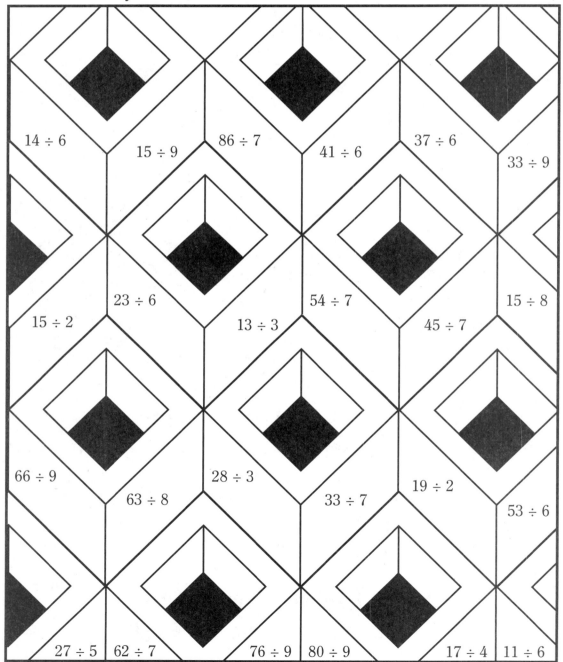

14 ÷ 6   15 ÷ 9   86 ÷ 7   41 ÷ 6   37 ÷ 6   33 ÷ 9

23 ÷ 6   15 ÷ 2   13 ÷ 3   54 ÷ 7   45 ÷ 7   15 ÷ 8

66 ÷ 9   63 ÷ 8   28 ÷ 3   33 ÷ 7   19 ÷ 2   53 ÷ 6

27 ÷ 5   62 ÷ 7   76 ÷ 9   80 ÷ 9   17 ÷ 4   11 ÷ 6

Division Decoder

Name _____ Date _____

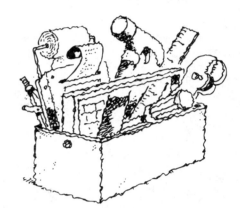

Riddle: What kind of tools do you use for math?

Find each quotient. Then use the Decoder to solve the riddle by filling in the spaces at the bottom of the page.

❶ $8 \div 2$ = _____

❷ $10 \div 5$ = _____

❸ $24 \div 4$ = _____

❹ $50 \div 10$ = _____

❺ $72 \div 9$ = _____

❻ $32 \div 10$ = _____

❼ $48 \div 7$ = _____

❽ $29 \div 3$ = _____

❾ $65 \div 8$ = _____

❿ $92 \div 6$ = _____

Decoder

8 .... **I**

3 remainder 2 .. **L**

7 .... **W**

8 remainder 1 .. **S**

6 .... **U**

9 .... **A**

15 remainder 3 . **B**

4 .... **L**

2 remainder 3 ... **D**

9 remainder 2 ... **T**

1 .... **F**

7 remainder 6 ... **N**

6 remainder 6 .... **I**

2 .... **E**

11 .... **O**

15 remainder 2 . **P**

2 remainder 5 ... **X**

10 .... **C**

5 .... **R**

"M ___ ___ ___ ___ " ___ ___ ___ ___ ___ ___
    3   1   8   5   10 6   7   2   4   9

Food Fractions

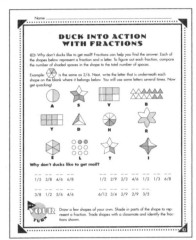

ACTIVITY GOAL

Identify the fraction represented in each shape to complete a riddle.

HELPFUL HINT!

• The denominator of each fraction represents the total amount of parts in the shape. The shaded parts represent the numerator. A food item can help illustrate this strategy.

EXAMPLE

There are 8 slices of pie shown here (/8) the denominator. The shaded area represents how many pieces of the pie you can eat (1/) the numerator. The fraction represented in this picture is 1/8.

TRY THIS!

Draw a pizza on another piece of paper, then cut out the circle. Cut the pizza into six equal pieces. Using your paper pizza, make the following fractions: 1/2, 2/6, 5/6, 1/3.
Now divide your pizza between you and two imaginary friends. Did you each get the same amount? _____

MORE SWEET FUN!

Color 1/3 of these 12 pieces of candy. What fraction of the candy is left? _____

Now that you've reviewed fractions, duck into action and name a few fractions to solve the riddle on the following page!

Duck Into Action With Fractions

Name _____ Date _____

✏ Why don't ducks like to get mail? Fractions can help you find the answer. Each of the shapes below represent a fraction and a letter. To figure out each fraction, compare the number of shaded spaces in the shape to the total number of spaces.

Example: ⬡ is the same as 2/6. Next, write the letter that is underneath each shape on the corresponding blank below. You will use some letters several times. Now get quacking!

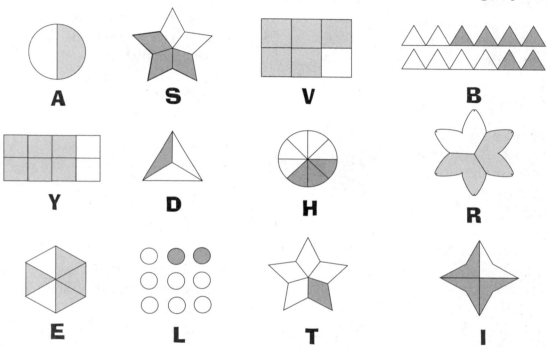

A   S   V   B

Y   D   H   R

E   L   T   I

Why don't ducks like to get mail?

___ ___ ___ ___   ___ ___ ___ ___ ___ ___ ___
1/5   3/8   4/6   6/8   1/2   2/9   2/3   4/6   1/2   1/3   6/8

___ ___ ___ ___   ___ ___ ___ ___ ___ .
3/8   1/2   5/6   4/6   6/12 3/4   2/9   2/9   3/5

Draw several shapes of your own. Shade in parts of the shape to represent a fraction. Trade shapes with a classmate and identify the fractions shown.

A Tasty Fraction Pie

Name _____ Date _____

Once, there lived a chef named Sheri. One morning, she emptied out all of her cupboards and made a magic pie—

Before you finish the story, make your own pie! Here's how.

You Need:

2 plain paper plates (each 8 inches) ◆ scissors ◆ crayons ◆ glue or paste

What to Do:

1. Color the pizza pattern from page 43 and paste it onto one plate. Cut along the dotted line, as shown.

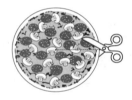

2. Mark the center point of the second plate.

3. Use a ruler to draw a straight line from the center of the second plate to its edge, like this.

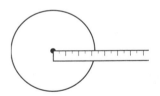

4. Cut along the lines on both plates.

5. Slip the two plates together at the slits, like this:

6. Now you have a magic fraction pie! And you are ready to go on with the story. Whenever you see a picture of Sheri's pie, make yours look the same.

The Amazing Fraction Pie

At noon, Sheri took her magic pie out of the oven. It was so pretty, she couldn't bear to cut it up. So Sheri put the whole pie on a table for everyone to see.

At 3:00, Sheri's cat, Sherlock, spotted the pie. To Sherlock, the pie looked tastier than meow chow! So he nibbled off $\frac{1}{4}$ of it. Now just $\frac{3}{4}$ of the magic pie was left.

At 6:00, Sheri's dog, Sam, sniffed the pie. It smelled better than a steak bone! So he put his paws on the table and took a big bite. Now only $\frac{1}{2}$ of the pie was left.

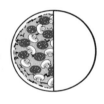

At 9:00, Sheri's friend Sarah stopped by for a snack. But Sheri was out shopping. So Sarah helped herself to a piece of the pie. Now $\frac{3}{4}$ of the pie was gone!

At 12:00, Sheri came home and ran to see her pretty pie. What do you think happened next? Write your own story ending here:

_____

_____

_____

_____

_____

A Tasty Fraction Pie

Into Infinity

Name _____ Date _____

Solve the problems. Then rename the answers in lowest terms.

If the answer is $\frac{1}{4}$, $\frac{1}{8}$, or $\frac{1}{16}$, color the shape purple.

If the answer is $\frac{1}{2}$, $\frac{1}{3}$, or $\frac{1}{7}$, color the shape blue.

If the answer is $\frac{2}{3}$, $\frac{3}{4}$, or $\frac{7}{8}$, color the shape green.

If the answer is $\frac{3}{5}$, $\frac{4}{5}$, or $\frac{5}{7}$, color the shape yellow.

If the answer is $\frac{9}{10}$ or $\frac{11}{12}$, color the shape red.

Finish the design by coloring the other shapes with colors of your choice.

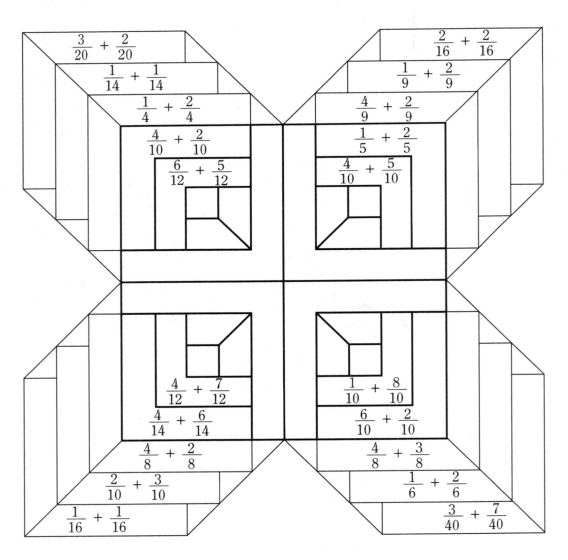

Trefoil

Name _____ Date _____

Solve the problems. ◆ Rename the answers in lowest terms. ◆ If the answer is $\frac{1}{2}$ or greater, color the shape red. ◆ If the answer is less than $\frac{1}{2}$, color the shape blue. ◆ Finish the design by coloring the other shapes with the colors of your choice.

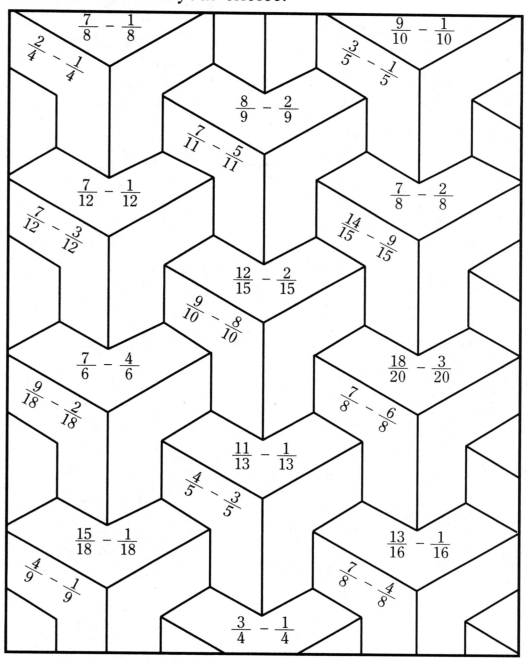

Kaleidoscope of Flowers

Name _____   Date _____

If the number has a 5 in the ones place, color the shape green.
If the number has a 5 in the tenths place, color the shape pink.
If the number has a 5 in the hundredths place, color the shape yellow.
Finish the design by coloring the other shapes with colors of your choice.

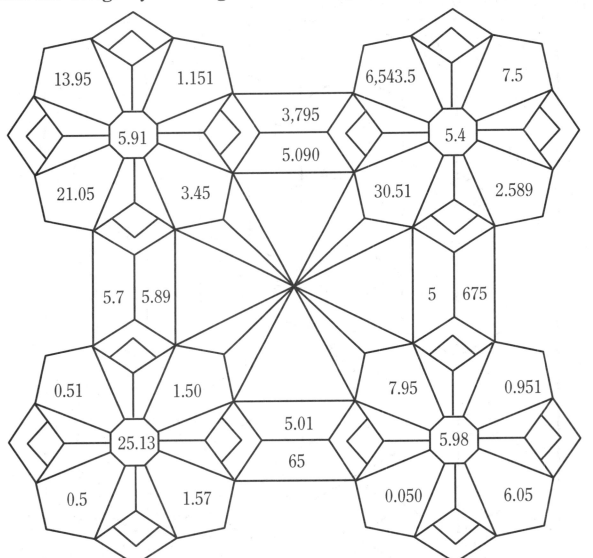

Taking It Further: Place the following decimals in the correct places on the lines below the dots: 4.9, 1.7, 2.5, and 0.2.

Decimals Around the Diamond

Name _____ Date _____

Baseball fans always argue who the best player was. Some say it was Ken Griffey, Jr. Others insist it was Cal Ripkin, Jr. Still others claim it was Barry Bonds. Everybody seems to have a favorite!

When it comes to finding the best hitter, though, no one can argue with batting averages. The batting average shows how often a baseball player gets a hit. It is a 3-digit decimal number, and looks like this: .328, .287, .311, .253. The larger the batting average is, the better the hitter is.

Decimals are numbers between 0 and 1. They are written to the right of the ones place. Decimals always have a decimal point to the left of them.

.325

decimal point — tenths place — hundredths place — thousandths place

Rank	Player (Team)	1995 Batting Average
	Cal Ripken, Jr. (Baltimore Orioles)	.262
	Barry Bonds (San Francisco Giants)	.294
	Mo Vaughn (Boston Red Sox)	.300
	Barry Larkin (Cincinnati Reds)	.319
	Kirby Puckett (Minnesota Twins)	.314
	Tony Gwynn (San Diego Padres)	.368
	Ken Griffey, Jr. (Seattle Mariners)	.258
	Mike Piazza (Los Angeles Dodgers)	.346
	Frank Thomas (Chicago White Sox)	.308
	David Justice (Atlanta Braves)	.253

What to Do:

Read the chart of baseball players' batting averages from 1995. Rank the batting averages. This means number the batting averages in order from highest to lowest. (See Home Plate for help.) Write the numbers 1 to 10 in the boxes next to the names—1 for the highest average, 10 for the lowest. Ready? Play ball!

HOME PLATE

To rank decimal numbers:
- Start at the left.
- Compare the digits in the same place.
- Find the first place where the digits are different.
- The number with the smaller digit is the smaller number. Example: Rank .317 and .312

.317
.312

So .312 is smaller than .317.

Across-and-Down Decimals

Name _____ Date _____

Complete the crossnumber puzzle as if it were a crossword puzzle.
Give each digit and decimal point its own square. Remember to align
the decimal points and add any necessary zeros, then proceed as if you
were adding whole numbers.

ACROSS		DOWN	
A.	1.3 + 2.4	A.	1.44 + 1.7
C.	2.2 + 2.18	B.	23.11 + 53.18
E.	.3 + .25	C.	2.25 + 2.25
F.	.3 + .3	D.	6.5 + 1.6
G.	.56 + .34	F.	.1604 + .11
J.	.4 + .17	H.	20.8 + 3.5
K.	6.93 + .23	I.	1.367 + .333
L.	1.18 + 3.12	J.	.2 + .16

Change Arranger

Name _____ Date _____

When you make change, always start with the price. Count on from the price. Start with the coins that have the least value. Write the change from these purchases.

1. LAWN GAME
AMOUNT GIVEN $5.00
PRICE 3.45
CHANGE $ _____

2. YO-YO
AMOUNT GIVEN $3.00
PRICE 2.77
CHANGE $ _____

3. BIKE HELMET
AMOUNT GIVEN $10.00
PRICE 7.55
CHANGE $ _____

4. SOAP BUBBLES
AMOUNT GIVEN $2.00
PRICE 1.52
CHANGE $ _____

5. VIDEO GAME
AMOUNT GIVEN $20.00
PRICE 7.30
CHANGE $ _____

6. ACTION TOY
AMOUNT GIVEN $10.00
PRICE 6.49
CHANGE $ _____

7. SUNGLASSES
AMOUNT GIVEN $4.00
PRICE 3.68
CHANGE $ _____

8. BACKPACK
AMOUNT GIVEN $20.00
PRICE 9.35
CHANGE $ _____

9. JUMP ROPE
AMOUNT GIVEN $4.00
PRICE 3.17
CHANGE $ _____

10. MARKERS
AMOUNT GIVEN $5.00
PRICE 2.43
CHANGE $ _____

Money Magic Puzzle

Name _____ Date _____

Round your answers to the nearest dollar. Circle the correct amount, then fill in the puzzle.

ACROSS:

A. $16.98 + $18.99 $36 $26

C. $24.85 + $29.99 $65 $55

E. $21.99 + $8.95 $31 $41

G. $218.04 + $67.90 $286 $386

I. $53.75 + $40.98 $105 $95

K. $7.99 + $19.70 $28 $22

L. $99.98 + 99.57 $300 $200

M. $65.75 + $20.90 $87 $97

O. $9.69 + $32.99 $40 $43

P. $588.95 + $14.90 $704 $604

Q. $3.75 + $9.99 $13 $14

R. $428.70 + $50.90 $480 $520

DOWN:

B. $28.59 + $33.95 $69 $63

C. $39.25 + $18.70 $58 $42

D. $376.35 + $184.50 $521 $561

F. $7.28 + $11.69 $19 $16

H. $199.80 + $224.99 $525 $425

J. $399.95 + $126.99 $527 $566

M. $5.85 + $76.95 $83 $75

N. $39.80 + $13.99 $54 $62

O. $26.98 + $16.89 $44 $49

P. $48.95 + $18.99 $68 $66

Time for Play

Name _____   Date _____

✏ The dogs in the neighborhood play in the park at the same time every day. Today, some are running around trees and others are playing catch with their owners. But most of them are busy doing something else—chasing another dog! What time were they chasing the dog? Equivalent measurements can help you find the answer.

DIRECTIONS:

• There are two answers next to each question. Circle the letter after the correct answer.
• When you've finished, write each circled letter in the blanks below the riddle. Be sure to write the letters in order.

1.	How many weeks are in a year?	34	**L**	52	**T**	
2.	How many inches are in a foot?	12	**W**	36	**A**	
3.	How many centimeters are in a meter?	100	**E**	1000	**O**	
4.	How many nickels are in a dollar?	40	**M**	20	**N**	
5.	How many days are in a year?	365	**T**	245	**S**	
6.	How many inches are in a yard?	36	**Y**	24	**B**	
7.	How many ounces are in a pound?	16	**A**	12	**I**	
8.	How many hours are in a day?	48	**C**	24	**F**	
9.	How many years are in a decade?	50	**H**	10	**T**	
10.	How many cups are in a pint?	2	**E**	4	**U**	
11.	How many quarts are in a gallon?	4	**R**	8	**D**	
12.	How many feet are in a mile?	5,280	**O**	2,160	**G**	
13.	How many seconds are in a minute?	30	**J**	60	**N**	
14.	How many millimeters are in a meter?	1,000	**E**	1500	**P**	

What time is it when twenty dogs run after one dog?

— — — — — — —   — — — —   — — — — —

Come up with an equivalent measurement problem of your own.

Measure by Measure

Name _____ Date _____

✏ Josie is surrounded by all kinds of measuring tools. But she's not sure which tool does what! Sure, she knows that a ruler measures the length of something. But she doesn't realize that all the other tools around her are used for measuring things too. Try giving Josie a hand.

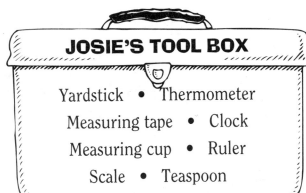

JOSIE'S TOOL BOX

Yardstick • Thermometer

Measuring tape • Clock

Measuring cup • Ruler

Scale • Teaspoon

DIRECTIONS:

Take a look at the list of measuring tools in Josie's Tool Box. Use the list to answer the questions below.

1. What tool could Josie use to measure the weight of a pumpkin? _____

2. What tool could Josie use to measure the width of her math book?

_____

3. Josie plans to watch one of her favorite television shows. What tool could help her measure the length of each commercial that appears during that show?

_____

4. Josie has an awful cough. What tool could she use to measure the amount of cough syrup she should take? _____

5. If Josie's mom wants to find out Josie's temperature, which tool could she use?

_____

6. Say Josie wanted to make a cake. What tool could she use to measure the milk she needs to put in the cake mix? _____

7. What tool could Josie use to measure the height of her brother's tree house?

_____

8. What tool could Josie give her dad to measure the length of their living room?

_____

IT'S YOUR TURN

Choose four of the measuring tools in Josie's Tool Box. Make a list of things you could measure with each of those tools.

Picnic Area

Name _____ Date _____

What to Do:

Area measures the number of square units inside a shape. Find the area of each ant family's picnic blanket by counting the number of squares on the blanket. Then answer the following questions.

Remember—area is measured in square units, such as square centimeters. My blanket's area is four square units.

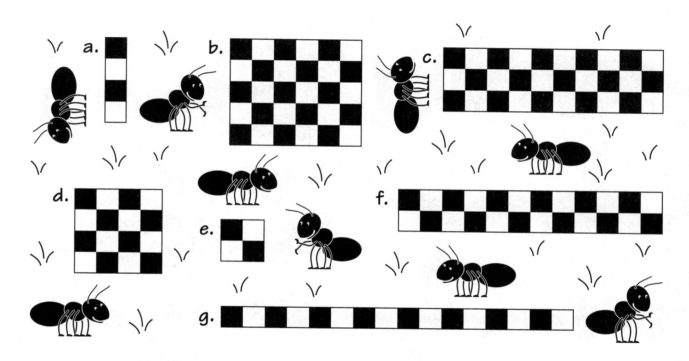

1. Which pairs of blankets have the same area?

_____ and _____

_____ and _____

_____ and _____

2. Which two blankets can you put together to make a rectangle with an area of 20?

_____

3. Which three blankets can you put together to make a rectangle with an area of 50?

_____

4. What is the total area of all of the ants' blankets?

_____

Perimeter and Area Zoo

Name _____   Date _____

A shape doesn't have to be a square or a rectangle to have perimeter and area. The animals in this zoo are different shapes. Can you find each animal's perimeter and area?

Remember: To find perimeter, count the sides of the units. To find area, count the number of whole units.

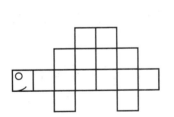

1. Perimeter _____

Area _____

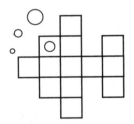

2. Perimeter _____

Area _____

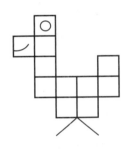

3. Perimeter _____

Area _____

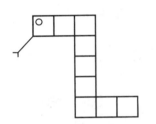

4. Perimeter _____

Area _____

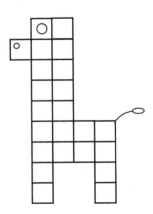

5. Perimeter _____

Area _____

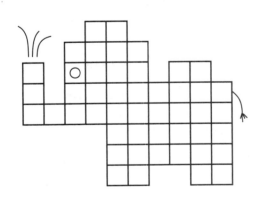

6. Perimeter _____

Area _____

Angles from A to Z

Name _____ Date _____

Angles are hiding everywhere—even in the words you're reading now. When two straight lines meet, they make an angle. There are three kinds of angles:

- The corner of a square or rectangle makes a right angle.

- Angles that are smaller than right angles are called acute angles.

- Angles that are larger than right angles are called obtuse angles.

 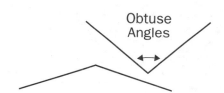

Take a look at the letters below. Circle each angle you see in the letters. Tell whether it is right, acute, or obtuse.

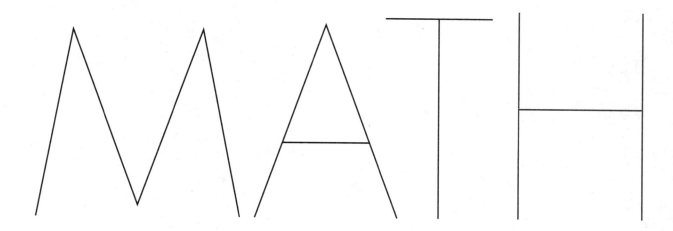

Flying Through the Air

Name _____   Date _____

What is the last thing that the trapeze flier wants to be?

Find the symmetrical shapes. Then use the Decoder to solve the riddle by filling in the blanks at the bottom of the page.

❶ ▢   _____

❷ ◿   _____

❸ ◖   _____

❹ ⚝   _____

❺   _____

❻   _____

❼   _____

❽   _____

❾   _____

❿   _____

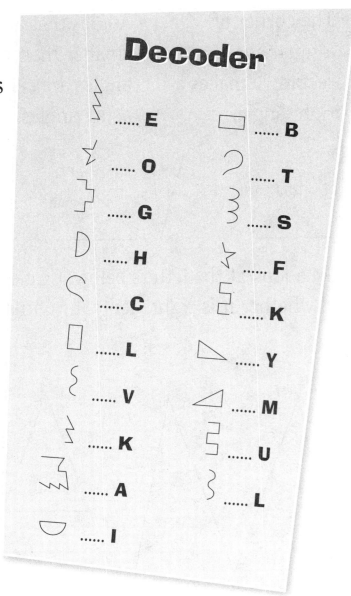

Decoder

...... E
...... O
...... G
...... H
...... C
...... L
...... V
...... K
...... A
...... I

...... B
...... T
...... S
...... F
...... K
...... Y
...... M
...... U
...... L

___ ___ ___ ___ ___ ___ ___ ___ ___ ___
6   3   9   4   10 1   7   5   8   2

Shape Up!

Name _____ Date _____

✏ How well do you know geometric shapes? Here's your chance to test yourself. Take a look at the shape in each statement. Fill in the blank spaces with the correct answers. When you're done, write the letters in the shaded squares on the spaces provided to solve the riddle.

What did the alien eat for lunch?

1. An ⬡ has __ ▢ __ __ ▢ sides.

2. This triangle has an angle that is the opposite of obtuse.

   It's an ▢ __ __ ▢ angle. ◺

3. The __ __ __ ▢ __ __ __ __ __ of this rectangle is fourteen.

4. The ▢ ▢ __ __ of this rectangle is twelve. ³⟦4⟧   ³⟦4⟧

5. This shape ▢ is a ▢ __ __ __ __ __.

6. This shape ⬛ is a __ __ ▢ __.

7. This shape △ is a __ __ __ __ ▢ __ __ __ __.

8. This shape ◯ is a __ __ ▢ __ __ __.

9. These shapes ⬡⬠⬡ have many sides.

   They are called __ __ __ __ __ __ __ ▢.

What did the alien eat for lunch?

__ __ __   __ __ __ __   __ __ __ __ __   __ __ __ __.

Draw a geometric shape not included in this activity on a piece of paper. Give it to a friend. See if he or she can name the shape.

Terrific Tessellations

Name _____ Date _____

What do math and art have in common? Everything—if you're making tessellations!

A **tessellation** (tess-uh-LAY-shun) is a design made of shapes that fit together like puzzle pieces. People use tessellations to decorate walls and floors, and even works of art.

This sidewalk is formed from rectangles.

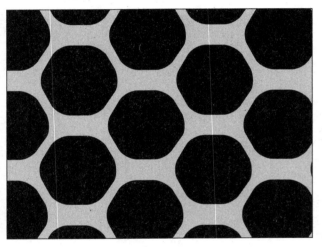

Hexagons form this beehive.

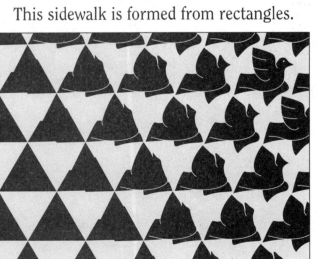

Here is a tessellation made from more than one shape.

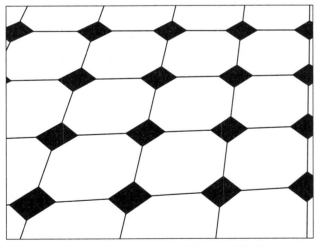

Squares and octagons form a tile floor.

Terrific Tessellations

Name _____ Date _____

You Need:
heavy paper ◆ scissors
tape ◆ crayons

What to Do:
Here's how you can make your own tessellation.

1. Start with a simple shape like a square. (Cut your shape from the heavy paper). Cut a piece out of side A . . .

2. . . . and slide it over to side B. Make sure it lines up evenly with the cut out side, or your tessellation won't work. Tape it in place on side B.

3. If you like, do the same thing with sides C and D. Now you have a new shape.

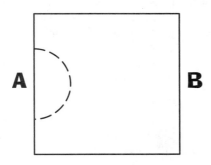

4. Trace your new shape on paper. Then slide the shape so it fits together with the one you just traced. Trace it again. Keep on sliding and tracing until your page is filled. Decorate your tessellation.

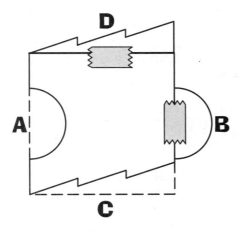

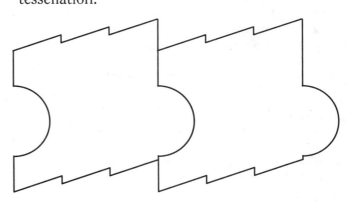

59

Answer Key

Page 5
1. 3,462; **2.** 98,724; **3.** 8,749
4. 128,467; **5.** 617,823; **6.** 618,397

Page 6
1. forty two; **2.** sixty eighth
3. six thousand two hundred thirty four; **4.** thirty four
5. one hundred forty five; **6.** seventeen
7. ninety three

Page 7
1. 496, 23; **2.** 49, 63; **or** 43, 69
3. 23, 496; **or** 26, 493; **or** 93, 426; **or** 96, 423
4. 732, 469
5. 23, 64, 97; **or** 23, 67, 94; **or** 24, 63, 97; **or** 24, 67, 93;
　　or 27, 63, 94; **or** 27, 64, 93
6. 674, 392

Page 8
1. 9,875; **2.** 12,346; **3.** 93,876
4. 13,579; **5.** 951,876; **6.** 123,648

Page 9

Page 10
1. 10; **2.** 20; **3.** 50; **4.** 90; **5.** 200
6. 400; **7.** 600; **8.** 300; **9.** 500; **10.** 700
What did the farmer get when he tried to reach the beehive?
A "buzzy" signal

Page 11
1. Exact Answer; **2.** Estimate; **3.** Exact Answer; **4.** Estimate
5. Estimate; **6.** Exact Answer; **7.** Estimate; **8.** Exact Answer
How Would You Estimate...
Answers will vary. Ask students to explain their estimation
method in their writing.

Page 12
1. 1; **2.** 4; **3.** 22; **4.** 26; **5.** 34
6. 57; **7.** 115; **8.** 124; **9.** 60; **10.** 215
How can you make the number seven even?
Take away the "s."

Page 13

Page 14
1. 16 apples; **2.** 11 apples; **3.** 40 apples; **4.** 6 apples; **5.** 14 apples
6. Jim; **7.** 17 oranges; **8.** 18 pears; **9.** 25 bananas; **10.** 45 apples
What is a ten without its number-one friend?
A total zero

Page 15
1. Deluxe Scraps
2. Kibble or Mouse Crumbs
3. Mouse Crumbs
4. Table Scraps and Mouse Crumbs
5. Crumbs & Cheese and Deluxe Scraps
Super Challenge: Woovis can't buy three items with $5. The
cheapest three items cost a combined $5.67.

Page 16
1. T; **2.** R; **3.** R; **4.** B; **5.** O
6. O; **7.** E; **8.** O; **9.** F; **10.** E
What do cheerleaders like to drink?
Lots of root beer

Page 17
1. bar 8; **2.** bar 6; **3.** 3 inches; **4.** 4 inches; **5.** 5 inches
6. 8 inches; **7.** bar 7; **8.** bar 3; **9.** bar 5; **10.** 4
Why did the horse sneeze?
It had a little "colt."

Page 18
1. $\frac{1}{2}$ white, $\frac{1}{2}$ black
2. $\frac{4}{7}$ black, $\frac{3}{7}$ white
3. $\frac{1}{4}$ black
4. 2 red socks, 4 green socks
5. $\frac{2}{5}$ red
Super Challenge: 5 yellow socks, 7 red and white socks

Page 19
19 + 37 = 56; 33 + 47 = 80; 51 + 39 = 90; 49 + 22 = 71
48 + 23 = 71; 78 + 12 = 90; 67 + 13 = 80; 29 + 27 = 56

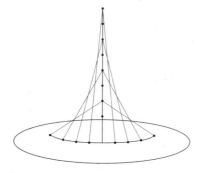

Taking It Further: a. 3; b. 9; c. 7; d. 85; e. 4; f. 26

Page 20
1. 19; **2.** 42; **3.** 54; **4.** 35; **5.** 66
6. 165; **7.** 245; **8.** 368; **9.** 768; **10.** 1,323
What sickness can't you talk about until it's cured?
Laryngitis

Page 21
102 + 49 = 151; 311 + 219 = 530; 315 + 669 = 984
452 + 266 = 718; 167 + 267 = 434; 202 + 179 = 381
177 + 114 = 291; 372 + 109 = 481; 345 + 166 = 511
201 + 99 = 300; 99 + 246 = 345; 397 + 119 = 516
245 + 345 = 590; 548 + 328 = 876; 756 + 145 = 901
219 + 79 = 298
Taking It Further: 750, 900, 1,050

Page 22
1. 1,804; **2.** 689; **3.** 1,063; **4.** 2,133; **5.** 3,489
6. 8,234; **7.** 7,538; **8.** 8,292; **9.** 5,429; **10.** 10,439
What's the best thing to eat in a bathtub?
Sponge cake

Page 23
Asteroids, 24–23
Comets, 98–96

Page 24
1. 6; **2.** 3; **3.** 28; **4.** 8; **5.** 18
6. 37; **7.** 97; **8.** 11; **9.** 64; **10.** 486
When is the ocean friendliest?
When it waves

Page 25
784 - 111 = 673; 598 - 247 = 351; 214 - 103 = 111
965 - 133 = 832; 999 - 399 = 600; 473 - 142 = 331
814 - 210 = 604; 692 - 581 = 111; 710 - 210 = 500
975 - 152 = 823; 388 - 221 = 167; 683 - 233 = 450
997 - 196 = 801; 678 - 333 = 345; 985 - 205 = 780
577 - 241 = 336; 988 - 117 = 871; 801 - 200 = 601
782 - 251 = 531; 890 - 730 = 160

Page 26
996 - 198 = 798; 756 - 267 = 489; 609 - 419 = 190
493 - 108 = 385; 815 - 124 = 691; 587 - 498 = 89
890 - 198 = 692; 752 - 653 = 99; 749 - 98 = 651
872 - 289 = 583; 707 - 498 = 209; 984 - 398 = 586
813 - 174 = 639

Page 27
45 x 22 = 990; 89 x 23 = 2047; 75 x 13 = 975; 68 x 26 = 1768
80 x 26 = 2080; 61 x 23 = 1403; 75 x 23 = 1725; 48 x 48 = 2304
96 x 42 = 4032; 77 x 23 = 1771; 49 x 71 = 3479; 24 x 20 = 480
96 x 26 = 2496; 65 x 13 = 845; 45 x 31 = 1395; 54 x 28 = 1512
34 x 28 = 952; 47 x 23 = 1081; 59 x 29 = 1711; 16 x 36 = 576
55 x 52 = 2860; 65 x 21 = 1365; 27 x 57 = 1539; 45 x 22 = 990
Taking It Further: a. 2; b. 1; c. 5

Page 28
57 x 73 = 4161; 98 x 34 = 3332; 23 x 13 = 299; 30 x 42 = 1260
21 x 61 = 1281; 44 x 20 = 880; 87 x 33 = 2871; 79 x 12 = 948
81 x 14 = 1134; 55 x 13 = 715; 58 x 42 = 2436; 25 x 13 = 325
60 x 33 = 1980; 61 x 11 = 671; 72 x 32 = 2304; 41 x 23 = 943
16 x 34 = 544; 53 x 73 = 3869; 27 x 34 = 918; 71 x 17 = 1207
49 x 52 = 2548; 83 x 17 = 1411; 25 x 46 = 1150; 95 x 36 = 3420
25 x 17 = 425; 62 x 12 = 744; 23 x 45 = 1035; 26 x 35 = 910
37 x 11 = 407; 24 x 20 = 480

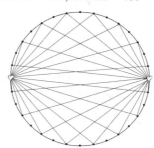

a. 9; b. 7; c. 2; d. 2

Page 29
1. 2,300; **2.** 3,400; **3.** 9,300; **4.** 17,600; **5.** 9,500
6. 16,200; **7.** 24,500; **8.** 14,400; **9.** 45,000
What are the cheapest ships to buy?
"Sale"boats

Page 30
973 x 8 = 7784; 380 x 6 = 2280; 909 x 7 = 6363; 178 x 4 = 712
272 x 8 = 2176; 319 x 4 = 1276; 592 x 7 = 4144; 711 x 7 = 4977
699 x 6 = 4194; 716 x 8 = 5728; 340 x 5 = 1700; 234 x 1 = 234
443 x 4 = 1772; 841 x 8 = 6728; 361 x 6 = 2166; 246 x 4 = 984
588 x 9 = 5292; 741 x 8 = 5928; 508 x 7 = 3556; 799 x 8 = 6392
247 x 7 = 1729; 300 x 9 = 2700; 650 x 9 = 5850; 348 x 9 = 3132
948 x 9 = 8532
Taking It Further: 222

Page 31
1. 11,000; **2.** 24,000; **3.** 30,000; **4.** 56,000; **5.** 100,000
6. 144,000; **7.** 210,000; **8.** 256,000; **9.** 360,000; **10.** 375,000
Why did the spider join the baseball team?
To catch "flies"

Page 32
14 ÷ 7 = 2; 28 ÷ 7 = 4; 63 ÷ 7 = 9; 49 ÷ 7 = 7
84 ÷ 7 = 12; 70 ÷ 7 = 10; 21 ÷ 7 = 3; 42 ÷ 7 = 6

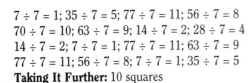
7 ÷ 7 = 1; 35 ÷ 7 = 5; 77 ÷ 7 = 11; 56 ÷ 7 = 8
70 ÷ 7 = 10; 63 ÷ 7 = 9; 14 ÷ 7 = 2; 28 ÷ 7 = 4
14 ÷ 7 = 2; 7 ÷ 7 = 1; 77 ÷ 7 = 11; 63 ÷ 7 = 9
77 ÷ 7 = 11; 56 ÷ 7 = 8; 7 ÷ 7 = 1; 35 ÷ 7 = 5
Taking It Further: 10 squares

Page 33
72 ÷ 6 = 12; 54 ÷ 6 = 9; 12 ÷ 6 = 2; 42 ÷ 6 = 7
36 ÷ 6 = 6; 60 ÷ 6 = 10; 48 ÷ 6 = 8; 18 ÷ 6 = 3
24 ÷ 6 = 4; 30 ÷ 6 = 5; 66 ÷ 6 = 11

Page 34
24 ÷ 6 = 4; 27 ÷ 3 = 9; 36 ÷ 6 = 6; 28 ÷ 4 = 7
32 ÷ 4 = 8; 40 ÷ 8 = 5; 7 ÷ 7 = 1; 18 ÷ 9 = 2
9 ÷ 9 = 1; 16 ÷ 8 = 2; 32 ÷ 4 = 8; 20 ÷ 4 = 5
42 ÷ 7 = 6; 63 ÷ 9 = 7; 16 ÷ 4 = 4; 81 ÷ 9 = 9

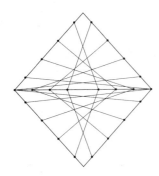

Taking It Further: 54, 45, 36, 27, 18; 42, 35, 28, 21, 14

Page 35
12 ÷ 3 = 4; 18 ÷ 3 = 6; 15 ÷ 5 = 3; 56 ÷ 7 = 8
18 ÷ 9 = 2; 27 ÷ 9 = 3; 40 ÷ 5 = 8; 16 ÷ 4 = 4
20 ÷ 5 = 4; 64 ÷ 8 = 8; 25 ÷ 5 = 5; 36 ÷ 6 = 6
10 ÷ 5 = 2; 32 ÷ 4 = 8; 42 ÷ 6 = 7; 36 ÷ 9 = 4
12 ÷ 4 = 3; 72 ÷ 9 = 8; 30 ÷ 5 = 6; 18 ÷ 9 = 2
49 ÷ 7 = 7

Page 36
4 ÷ 4 = 1; 28 ÷ 4 = 7; 36 ÷ 4 = 9; 16 ÷ 4 = 4
8 ÷ 4 = 2; 32 ÷ 4 = 8; 20 ÷ 4 = 5; 12 ÷ 4 = 3
40 ÷ 5 = 8; 20 ÷ 5 = 4; 5 ÷ 5 = 1; 30 ÷ 5 = 6
15 ÷ 5 = 3; 10 ÷ 5 = 2; 35 ÷ 5 = 7; 45 ÷ 5 = 9

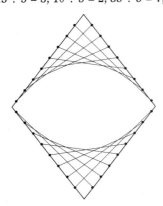

Taking It Further: 10

Page 37
880 ÷ 2 = 440; 996 ÷ 3 = 332; 576 ÷ 4 = 144; 502 ÷ 2 = 251
992 ÷ 2 = 496; 603 ÷ 3 = 201; 903 ÷ 3 = 301; 392 ÷ 2 = 196
982 ÷ 2 = 491; 897 ÷ 3 = 299; 738 ÷ 6 = 123; 742 ÷ 2 = 371
990 ÷ 3 = 330
Taking It Further:

```
        4 3 2
  2  )  8 6 4
    -   8
        0 6
      -   6
          0 4
        -   4
            0
```

Page 38
14 ÷ 6 = 2R2; 15 ÷ 9 = 1R6; 86 ÷ 7 = 12R2; 41 ÷ 6 = 6R5
37 ÷ 6 = 6R1; 33 ÷ 9 = 3R6; 15 ÷ 2 = 7R1; 23 ÷ 6 = 3R5
13 ÷ 3 = 4R1; 54 ÷ 7 = 7R5; 45 ÷ 7 = 6R3; 15 ÷ 8 = 1R7
66 ÷ 9 = 7R3; 63 ÷ 8 = 7R7; 28 ÷ 3 = 9R1; 33 ÷ 7 = 5R5
19 ÷ 2 = 9R1; 53 ÷ 6 = 8R5; 27 ÷ 5 = 5R2; 62 ÷ 7 = 8R6
76 ÷ 9 = 8R4; 80 ÷ 9 = 8R8; 17 ÷ 4 = 4R1; 11 ÷ 6 = 1R5

Page 39
1. 4; **2.** 2; **3.** 6; **4.** 5; **5.** 8; **6.** 3 remainder 2
7. 6 remainder 6; **8.** 9 remainder 2
9. 8 remainder 1; **10.** 15 remainder 2
What kind of tools do you use for math?
"Multi"pliers

Pages 40-41
T   H   E   Y   A   L   R   E   A   D   Y
1/5 3/8 4/6 6/8 1/2 2/9 2/3 4/6 1/2 1/3 6/8

H   A   V   E   B   I   L   L   S.
3/8 1/2 5/6 4/6 6/12 3/4 2/9 2/9 3/5

Pages 42-43
Children's stories will vary.

Page 44
3/20 + 2/20 = 1/4; 2/16 + 2/16 = 1/4; 1/14 + 1/14 = 1/7
1/9 + 2/9 = 1/3; 1/4 + 2/4 = 3/4; 4/9 + 2/9 = 2/3
4/10 + 2/10 = 3/5; 1/5 + 2/5 = 3/5; 6/12 + 5/12 = 11/12
4/10 + 5/10 = 9/10; 4/12 + 7/12 = 11/12; 1/10 + 8/10 = 9/10
4/14 + 6/14 = 5/7; 6/10 + 2/10 = 4/5; 4/8 + 2/8 = 3/4
4/8 + 3/8 = 7/8; 2/10 + 3/10 = 1/2; 1/6 + 2/6 = 1/2
1/16 + 1/16 = 1/8; 3/40 + 7/40 = 1/4

Page 45
7/8 - 1/8 = 3/4; 9/10 - 1/10 = 4/5; 2/4 - 1/4 = 1/4
3/5 - 1/5 = 2/5; 8/9 - 2/9 = 2/3; 7/11 - 5/11 = 2/11
7/12 - 1/12 = 1/2; 7/8 - 2/8 = 5/8; 7/12 - 3/12 = 1/3
14/15 - 9/15 = 1/3; 12/15 - 2/15 = 2/3; 9/10 - 8/10 = 1/10
7/6 - 4/6 = 1/2; 18/20 - 3/20 = 3/4; 9/18 - 2/18 = 7/18
7/8 - 6/8 = 1/8; 11/13 - 1/13 = 12/13; 4/5 - 3/5 = 1/5
15/18 - 1/18 = 7/9; 13/16 - 1/16 = 3/4; 4/9 - 1/9 = 1/3
7/8 - 4/8 = 3/8; 3/4 - 1/4 = 1/2

Page 46
Taking It Further:

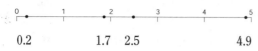

0.2   1.7   2.5   4.9

Page 47
Here is the correct ranking from highest to lowest:
1—Tony Gwynn (.368); 2—Mike Piazza (.346); 3—Barry Larkin (.319); 4—Kirby Puckett (.314); 5—Frank Thomas (.308); 6—Mo Vaughn (.300); 7—Barry Bonds (.294); 8—Cal Ripken, Jr. (.262); 9—Ken Griffey, Jr. (.258); 10—David Justice (.253)

Page 48
ACROSS: A. 3.7; **C.** 4.38; **E.** .55; **F.** .6; **G.** .9; **J.** .57 **K.** 7.16; **L.** 4.3
DOWN: A. 3.14; **B.** 78.29; **C.** 4.5; **D.** 8.1; **F.** .2704 **H.** 24.3; **I.** 1.7; **J.** .36

Page 49
1. $1.55; **2.** 23¢; **3.** $2.45; **4.** 48¢; **5.** $12.70
6. $3.51; **7.** 32¢; **8.** $10.65; **9.** 83¢; **10.** $2.57

Page 50

ᴬ3	ᴮ6			ᶜ5	ᴰ5
	ᴱ3	ᶠ1		ᴳ2 8	6
ᴴ4		ᴵ9	ᴶ5		1
ᴷ2	8		ᴸ2	0	0
5		ᴹ8	7		ᴺ5
	ᴼ4	3		ᴾ6	0 4
ᵠ1	4		ᴿ4	8	0

Page 51
1. 52; **2.** 12; **3.** 100; **4.** 20; **5.** 365; **6.** 36; **7.** 16
8. 24; **9.** 10; **10.** 2; **11.** 4; **12.** 5,280; **13.** 60; **14.** 1,000
Answer: TWENTY AFTER ONE

Page 52
1. scale; **2.** ruler; **3.** clock; **4.** teaspoon; **5.** thermometer
6. measuring cup; **7.** yardstick; **8.** measuring tape

Page 53
1. a and e (4); b and c (30); d and g (16)
2. a and d; **3.** a, b, and d; **4.** 124

Page 54
1. perimeter: 20; area: 9; **2.** perimeter: 24; area: 13
3. perimeter: 22; area: 11; **4.** perimeter: 24; area: 15
5. perimeter: 32; area: 23; **6.** perimeter: 44; area: 48

Page 55
1. eight; **2.** acute; **3.** perimeter; **4.** area
5. square; **6.** cube; **7.** triangle; **8.** circle; **9.** polygons
Answer: IT ATE MARS BARS.

Page 56
1.   6.
2.   7.
3.   8.
4.   9.
5.   10.

What is the last thing that the trapeze flier wants to be?
The fall guy

Page 57
Students can examine each angle in the letters MATH and determine whether it is right, acute, or obtuse.

Pages 58-59
Children's Tessellate patterns will vary.

Instant Skills Index